Report no. VNTSC-TFM-11-2

New Ways of Looking at Sector Demand and Sector Alerts

Eugene Gilbo
Scott Smith

March 2011

Prepared for
Federal Aviation Administration
Office of System Operations Programs
800 Independence Ave., SW
Washington, DC 20591

Prepared by
Volpe National Transportation Systems Center
Traffic Flow Management Division
55 Broadway
Cambridge, MA 02142

Acknowledgments

Within the FAA, we would like to thank John McCarron and Mike McKinney who have supported and encouraged this work, Tom St. Clair, Mark Holben, Bill Smayda, and Tanya Yuditsky for their valuable comments and suggestions. Within the Volpe Center, we would like to thank Jim Hill and Rick Oiesen for reviewing the report and their valuable comments and suggestions that improved its content. We also thank Mary Costello for editing this report.

Executive Summary

The problem. This report presents and analyzes new approaches for determining when a sector should be alerted by the Monitor Alert capability of the Traffic Flow Management System (TFMS). TFMS currently alerts a fifteen-minute interval for a sector if the demand for any minute in that interval exceeds the Monitor Alert Parameter (MAP). This method suffers from three problems:

- The demand from a single minute does not adequately reflect controllers' workload for the entire 15-minute interval.
- The alert status for an interval is unstable: it can frequently change with minute-by-minute TFMS updates.
- The alerted interval depends on the arbitrary 15-minute boundaries.

To deal with these problems, this report describes two new approaches for determining when a sector should be alerted, a deterministic demand pattern approach and a probabilistic approach.

Deterministic demand pattern approach. The heart of this approach is that instead of using the TFMS demand for a single minute to determine if a sector is alerted, it uses a pattern of demand over many minutes. It is not connected to 15-minute intervals. This approach is called the **demand pattern approach**, which is based on the patterns of overloaded and non-overloaded one-minute sector demands, where a minute is overloaded if the predicted demand for that minute exceeds the MAP. The patterns are defined by two parameters: the minimum number of overloaded minutes (not necessarily consecutive) sufficient for declaring a sector alert (the "on" parameter), and the minimum number of consecutive non-overloaded minutes sufficient to reset the alert (the "off" parameter). Although this approach is called deterministic and the patterns are based on TFMS deterministic predictions of one-minute sector demands, by using a pattern of demand instead of the demand for a single minute, the uncertain nature of the predicted demand is acknowledged and implicitly taken into account, and the problems described above are addressed.

Probabilistic approach. Two approaches are used that explicitly reflect the probabilistic nature of demand predictions. The **first** one, which is called **the weighted average approach**, is based on expected values of one-minute traffic demand, instead of one-minute counts predicted by TFMS, for sector demand predictions and identification of sector congestion in Monitor/Alert. The expected one-minute demand is determined as a weighted average of the minute-by-minute demands surrounding the minute of interest. This approach acknowledges that most flights will arrive somewhat earlier or somewhat later than the predicted one-minute interval, and the weights in the weighted average are chosen to reflect the probabilities of such occurrences.

The **second** approach, which is called a **fully probabilistic approach**, gives for each sector the probability that demand will exceed MAP. If desired, an alert could be declared if the probability of excess demand exceeds a specified percentage. This approach uses both the expected one-minute demand (via weighted average) and the corridor of uncertainty around the expected demand that determines the probabilities for predicted demand to be within or outside the corridor. Previous work done by Volpe developed the algorithms for calculating the probabilistic characteristics of sector demand predictions. In this work, it was shown how to derive the probability distribution for a flight's sector entry time and for its sector occupancy time and also how to use these flight-by-flight probability distributions to derive a probability distribution for the aggregate one-minute demand for a sector.

The analysis. The analysis is based on TFMS demand prediction data at several en route sectors during several days of 2009 and 2010. Software was written that processed the voluminous TFMS data and that calculated the results for the approaches described above.

Results for the demand patterns and the weighted average approaches.

- The demand patterns approach and the weighted average approach significantly reduced the number of alerted periods in comparison with current TFMS Monitor/Alert.

- The demand patterns approach and the weighted average approach significantly reduced the total duration of alerted periods in comparison with current TFMS Monitor/ Alert.

- The demand patterns approach reduced not only the total duration but also the average number of minutes per alert (or average duration of a single alert).

- The weighted average approach did not noticeably reduce the average number of minutes per alert but did significantly reduce the total duration of alerts in comparison with current TFMS Monitor/Alert.

- As the "on" parameter of demand patterns increases, it becomes harder to turn alerts on because of the increased minimum duration of an alerted period. There are fewer alerts and the total duration of alerts decreases.

- As the "off" parameter of demand patterns increases, it becomes harder to turn an alert off. As a result, the total duration of alerted periods tend to be higher with a higher "off" parameter.

- The number of alerted periods is significantly more sensitive to "on" parameter (under various "off" parameters) than to "off" parameter (under the same "on" parameter) of demand patterns.

Results for the fully probabilistic approach.

- Expected one-minute traffic demand predictions were estimated via a weighted average of several consecutive TFMS predictions. The weighted average was based on probabilistic considerations, namely the observed error distribution of sector entry times for active and proposed flights.

- We estimated a range of uncertainty in the one-minute demand predictions, and constructed uncertainty bands around the weighted average, also based on prediction errors for active and proposed flights.

- Based on weighted average and range of uncertainty, the probabilities of alerts were determined at various times, which would provide further information for TFM decision-making.

Conclusion. This report presents new approaches for improving the way that TFMS determines if a sector is alerted. The deterministic demand pattern approach as well as the probabilistic weighted average approach have the advantage of being a smaller change from the traditional and long accepted way of doing business, while the fully probabilistic approach has the advantage of providing a more satisfactory way of looking at demand. The FAA can now compare these approaches to the current approach and determine which is more attractive for the long run. If one of the new approaches is deemed desirable, additional work would be needed to flesh the proposals out into fully operational tools.

Table of Contents

1. Introduction 1
2. One-minute Demand Patterns vs. One-minute Peak for Alerting Sectors 4
3. Probabilistic Traffic Demand and Alerts 11
4. Analysis of TFMS One-minute Demand Predictions for En Route Sectors 17
5. Summary and Next Steps 28
6. References 31

1. Introduction

This report presents the latest results of research conducted at the Volpe Center on improving air traffic demand predictions and enhancing the Traffic Flow Management System (TFMS) Monitor/Alert function for identifying potential congestion at National Airspace System (NAS) elements such as airports, sectors and fixes. This research contributes to the FAA's effort to improve aviation safety and efficiency of utilizing the NAS operational resources while accommodating increased air traffic demand, which are among the major goals of the NextGen program [1], [2]. One of the means to increase aviation safety is improvement of strategic air Traffic Flow Management decision making, which requires more accurate and reliable predictions of congestion in the NAS. Reducing uncertainty in traffic demand and capacity predictions through application of probabilistic methods to TFM decision making are necessary research components to support NextGen goals.

This report deals with uncertainty in traffic demand predictions in en route sectors and with the identification of potential congestion in airspace. Currently, the TFMS measures demand and determines alert status at each 15-minute interval of the time period of interest by comparing demand with the Monitor/Alert Parameter (MAP). However, unlike airports and fixes, where demand is determined by aggregating flights in 15-minute intervals, the TFMS predicts sector demands for each minute of a 15-minute interval and uses the demand of the peak minute as the sector demand for entire 15-minute interval. In other words, the entire 15-minute interval is alerted even if predicted demand exceeds the MAP only in one of fifteen minutes. Using the peak demand from a single minute to declare sector alerts for entire 15-minute interval creates some problems for and has been criticized by many Traffic Flow Management (TFM) and Air traffic Control (ATC) specialists mostly because of low accuracy and reliability of TFMS sector Monitor/Alert functionality.

Problems with the current way of alerting sectors include the following:

- First, measuring sector demand for a 15-minute interval by a peak one-minute demand does not realistically represent sector demand for the whole 15-minute interval and **does not adequately reflect sector controllers' workload** for the interval.

- Second, reliance on a single one-minute peak demand causes **instability** of Monitor/Alert by making the alert status very sensitive to errors in demand predictions that sometimes results in flickering of alerts.

- Third, there is an element of **arbitrariness** in TFMS alerting 15-minute intervals. In particular, TFMS determines sector alert status in **15-minute intervals on the time line with the artificially assigned boundaries**, so that that either entire 15-minute interval is alerted or none of it is.

Several previous Volpe reports addressed these problems and proposed new ways of measuring sector demand for a better and more realistic identification of sector alert status.

The research was undertaken with two objectives:

- Increasing accuracy of and reducing uncertainty in sector one-minute demand counts predictions and

- Improving TFMS monitor/alert functionality.

In our 2005 report [3], the accuracy of TFMS demand predictions at airports and sectors was analyzed and a linear regression approach was proposed for reducing uncertainty and increasing accuracy in traffic demand predictions. The idea was to use a weighted average of TFMS predictions of traffic demand for several consecutive 15-minute intervals within a moving time window to estimate the traffic demand for the interval in the middle of the window. In other words, the algorithm for estimation of traffic demand in a specific 15-minute interval used TFMS demand predictions for several adjacent intervals, both preceding and following the interval of interest in a linear regression. Analysis showed that it was sufficient to use predictions in three consecutive 15-minute intervals (within a moving 45-minute time window) to achieve the main effect for improving accuracy of demand predictions. It was also shown that linear regression was more efficient for airports than for sectors, presumably because the demand metric is different for airports and sectors. It was more helpful to look at aggregate demand counts in adjacent intervals (for airports) than to look at peak demands in adjacent intervals (for sectors). As a result, the next research efforts were concentrated on sector demand and sector monitor/alert improvements.

In a 2008 report [4], a new concept of determining alert status was proposed based on the patterns of overloaded and non-overloaded (normal) one-minute sector demands. The concept received a positive response from several TFM specialists. The patterns are defined by two parameters: the minimum number of overloaded minutes (not necessarily consecutive) sufficient for declaring a sector alert (the "on" parameter), and the minimum number of consecutive non-overloaded minutes sufficient to reset the alert (the "off" parameter). Although this approach is called deterministic and the patterns are based on TFMS deterministic predictions of one-minute sector demands, by using a pattern of demand instead of the demand for a single minute, the uncertain nature of the predicted demand is acknowledged and implicitly taken into account, and the problems described above are addressed.

Characteristics of uncertainty in flight events predictions by ETMS (now TFMS), including the characteristics of errors in predicting sector entry times for individual flights, are presented in our 2008 report [5]. Those errors are a major contributor to low accuracy of sector one-minute demand predictions.

The report of January 2010 [6] addressed the uncertainty in sector demand predictions caused by random errors in estimated time of arrival (ETA) to a sector for the individual flights (Here, the ETA is the time of crossing the sector boundary.). During the study, the analytical method was developed for a probabilistic representation of one-minute traffic demand counts for both entering a sector and being in a sector. The probabilistic sector demand included the expected one-minute counts and the uncertainty area around the expected values where the traffic demand counts could be predicted with a certain probability range. An important practical result of the study was that the expected one-minute demand counts in a sector are equal to a weighted average of the minute-by-minute demand predictions within a sliding time window that contains a one-minute interval of interest surrounded by several immediate preceding and following one-minute intervals. This result acknowledges that most flights will arrive somewhat earlier or somewhat later than the predicted one-minute interval, and the weights in the weighted average are chosen to reflect the probabilities of such occurrences. The analytical results presented in [6] make it possible to determine both the width of the time window and the weight coefficients via the probability distribution of errors in predicting of ETAs for individual flights.

This analysis also provided an estimate of the standard deviation of the one minute demand counts. The distributions and their parameters depend on both flight status (active or proposed) and a look-ahead time (LAT). The explicit analytical expressions for expected values and standard deviations of one-minute demand counts were obtained that take into account characteristics of errors in ETA predictions for both active and proposed flights.

This report uses the results of our previous research and presents a comparative analysis of various options for alerting sectors while dealing with uncertainty in sector demand predictions. The options include both demand patterns consisting of TFMS one-minute deterministic demand predictions and probabilistic sector demands that are used for probabilistic predictions of sector congestion by the Monitor/Alert function.

The report is organized as follows:

- Section 2 explains the demand pattern concept for alerting sectors and illustrates the connections between parameters of the demand patterns and the periods of sector alerts.

- Section 3 describes and illustrates deterministic and probabilistic predictions of sector demands.

- Section 4 presents the results of statistical analysis of various demand patterns applied for alerting sectors. The demand patterns were tested on TFMS data for 16 en route sectors.
- Section 5 summarizes the results of the study and formulates next steps.

2. One-minute Demand Patterns vs. One-minute Peak for Alerting Sectors

Start with the current TFMS sector alert rule: a sector is alerted for a whole 15-minute interval if the peak one-minute demand count within the 15-minute interval exceeds a sector MAP.

Figure 2-1 illustrates the case when there is only one overloaded minute in each of two 15-minute intervals from 1200 to 1230.

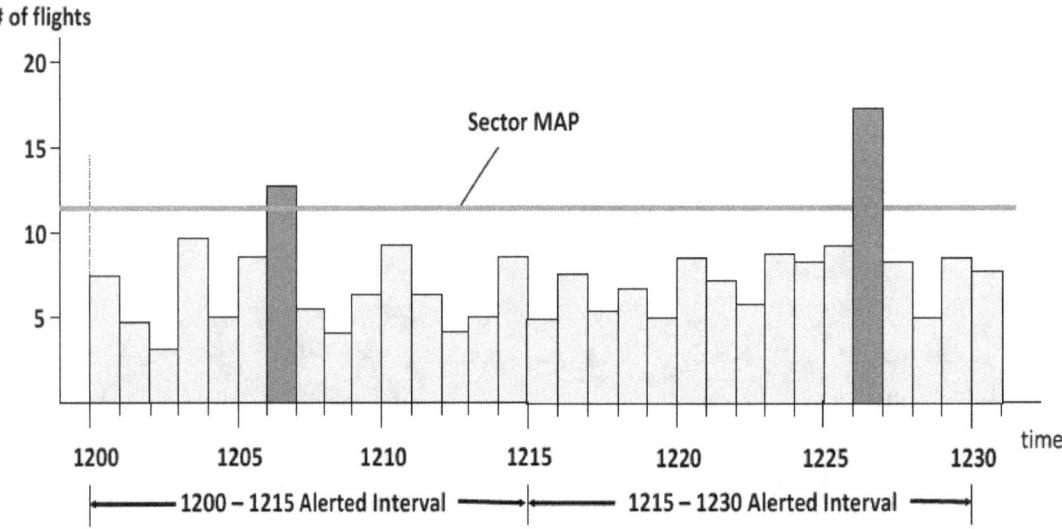

Figure 2-1 Example of Current TFMS Sector Alert Rule

In this example, TFMS would alert both 15-minute intervals, i.e., it would alert the entire 30-minute time period.

Figure 2-1 also illustrates the potential instability and inaccuracy of a sector alert. The excess demand for 1206 minute does not appear significant, and, due to random prediction errors, the next traffic demand update could be below MAP for the 1206 minute. If the other updated minutes in the first 15-minute interval remain normal, the entire 1200 – 1215 interval would not be alerted by TFMS. The next update, however, could return the 1200 – 1215 interval back to alerted status.

This simple example illustrates the potential problems with current TFMS sector Monitor/Alert concerning workload, instability, and arbitrariness that were mentioned in the introduction to this report.

The detailed motivation for, and explanation of, a new approach to alerting sectors based on one-minute demand patterns can be found in the Volpe Center's 2008 report [4]. Here, we present an overview of one-minute demand patterns that can be used for identifying sector alerts.

A demand pattern is a combination of one-minute intervals where traffic demands exceed sector MAP in some or all of the intervals. Any minute for which the predicted demand exceeds the MAP is called an overloaded minute. Otherwise a minute will be called a non-overloaded or a normal minute. Hence, the sector demand pattern is a combination of overloaded and normal minutes.

To identify a sector alert, two parameters of demand patterns are introduced:

- parameter *a* that determines a minimum number of overloaded minutes (not necessarily consecutive) sufficient for declaring an alert
- parameter *b* that indicates a minimum number of consecutive normal minutes between two overloaded minutes required to "reset" an alert, i.e., whenever **b** or more consecutive normal minutes are encountered, the counting for identification of next alerted interval will be restarted at the end of the string of normal minutes. This parameter characterizes a density of overloaded minutes necessary for declaring alerts and helps determine the start and the end of an alerted interval.

To illustrate how these alerting rules would work, consider the example pictured in Figure 2-2. This figure shows predicted sector demand at each minute of the 31-minute period that starts at 1200 and runs through 1230. Assume **a** = 3, **b** = 3. This means that a sector alert is declared for any interval with at least three overloaded minutes and with less than three consecutive normal minutes.

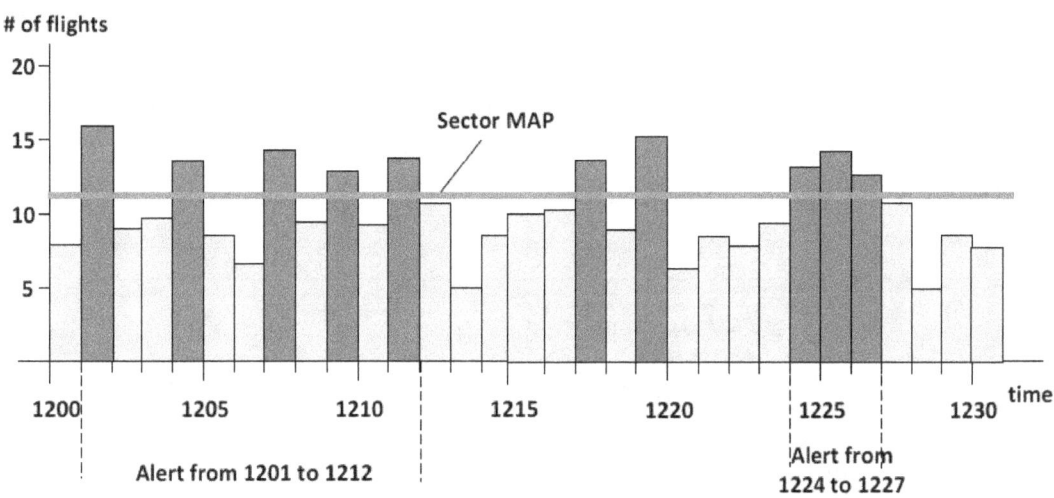

Figure 2-2 Example of a Pattern of Demand that Results in Two Alerted Intervals

In Figure 2-2, the checking for sector alert starts from the first overloaded 1201 minute. Just after 1201, there are two consecutive normal minutes followed by one overloaded 1204 minute. (So far, we have two overloaded minutes separated by less than three consecutive normal minutes, so that we can continue count). The next overloaded 1207 minute is also separated from the previous overloaded 1204 minute by less than three consecutive normal minutes. Thus, there are three overloaded minutes, close to each other, which is enough for declaring a sector alert that starts at 1201. We need to continue counting overloaded and normal minutes to find the end time for sector alert. Figure 2-2 shows that the three subsequent overloaded minutes are separated by less than three (actually by one) normal minutes, and the 1211 overloaded minute is followed by five (more than three) normal minutes. Hence, the alerted interval starts at 1201 and ends at 1212. The search for a candidate alert interval should then be restarted until encountering to the first overloaded 1217 minute. The next overloaded 1219 minute is separated from the 1217 overloaded minute by one normal minute so that there are two closely separated overloaded minutes, and this is not enough so far for alerting a sector. After the 1219 overloaded minute, there are four normal minutes. It means that interval 1217 – 1220, containing only two overloaded minutes, cannot be alerted, and the search for alerted interval should restart and continue. The next alerted interval is 1224-1227, because it contains three consecutive overloaded minutes followed by four normal minutes.

It is worth noting that in the pictured interval there are two alerted intervals lasting 3 and 11 minutes for a total of 14 alerted minutes; in contrast, TFMS currently would alert two consecutive 15-minute intervals, i.e., it would alert for 30 minutes from 1200 to 1230.

Looking not at a single minute but at the pattern of demand over time holds out the possibility of dealing with the problems with the current TFMS concept that were discussed in the Introduction. First, the pattern that constitutes an alert can be chosen to try to capture scenarios that represent real traffic management problems and better reflect controller's workload. Second, since the pattern takes into account multiple minutes rather than just one, it improves accuracy and stability of alerts. Third, the pattern is not affected by the boundaries of 15-minute intervals.

Below, we will present some examples based on TFMS sector demand data to illustrate how demand patterns affect the alerted intervals.

For identification of demand patterns, the following notation will be used: **a** on, **b** off. It means that **a** or more overloaded minutes turn alert on, and **b** or more consecutive non-overloaded minutes turn alert off.

Figure 2-3, which is taken from actual minute-by-minute predictions, shows traffic demand predicted for each one-minute interval of a 2hr 15min (or 135min) period in a sector with the sector MAP = 15. The boundary of the first 15-minute interval starts at LAT = 0. (LAT stands for look-ahead time. This refers to how far in the future a prediction is being made.) Each bar in the figure shows active and proposed fractions of one-minute demands by different colors.

In the figure, there are several overloaded minutes, where one-minute demands exceed the sector MAP.

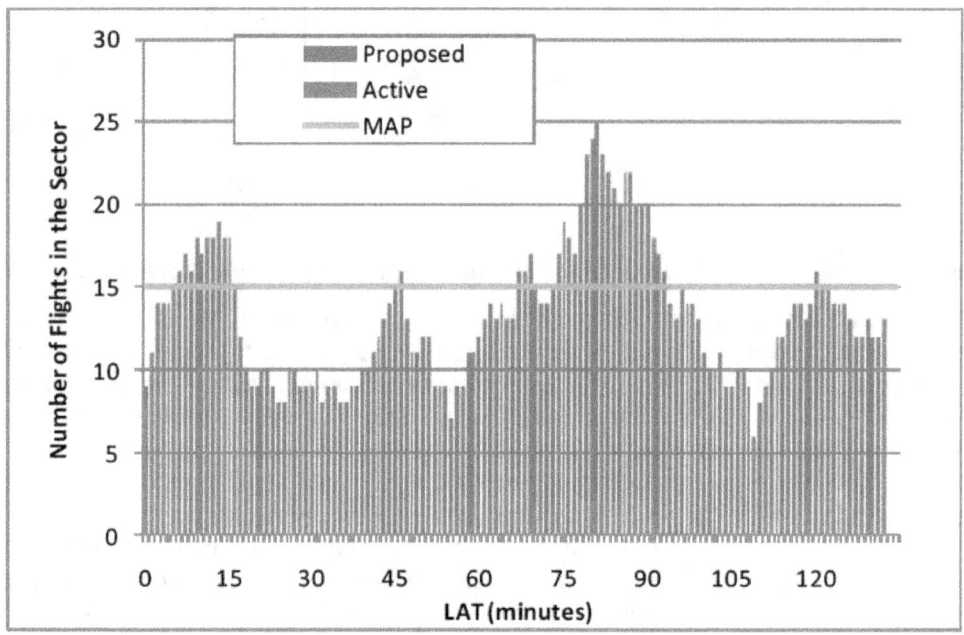

Figure 2-3 Predicted Traffic Demand

Of nine 15-minute intervals comprising the 135-minute period, current TFMS Monitor/Alert would alert all but two 15-minute intervals, because most of the intervals have at least one overloaded minute. Figure 2-4 highlights the alerted intervals.

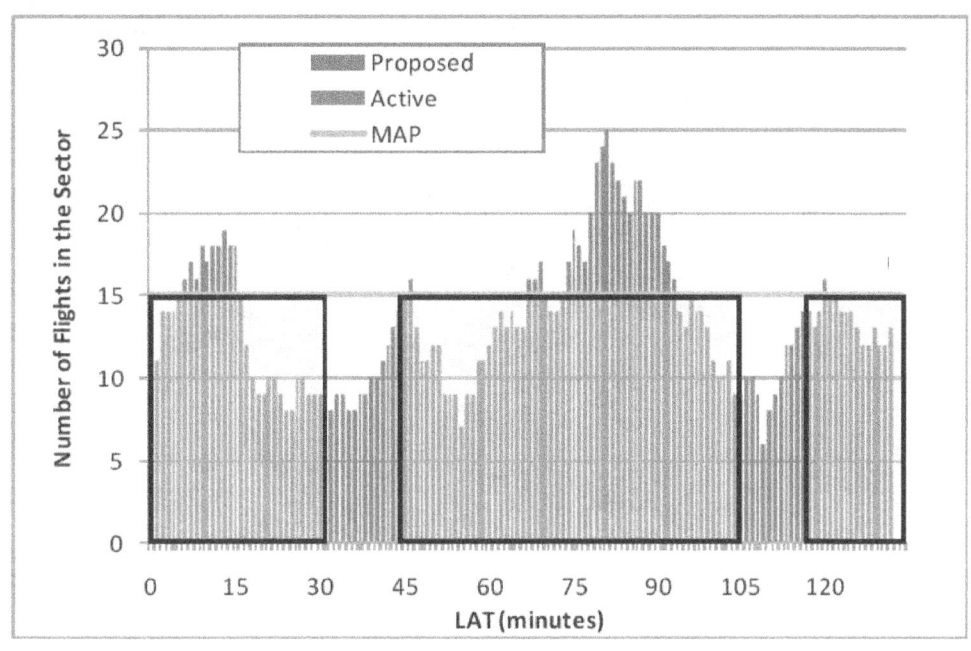

Figure 2-4 Sector Alert Periods Identified by Current TFMS

Subsequent illustrations will focus on a 60-minute fraction of sector demand predictions with LAT between 45 and 105, which represents the large central peak in Figures 2-3 and 2-4.

The "3 on, 3 off" sector alerting rule, applied to this demand, is shown in Figure 2-5.

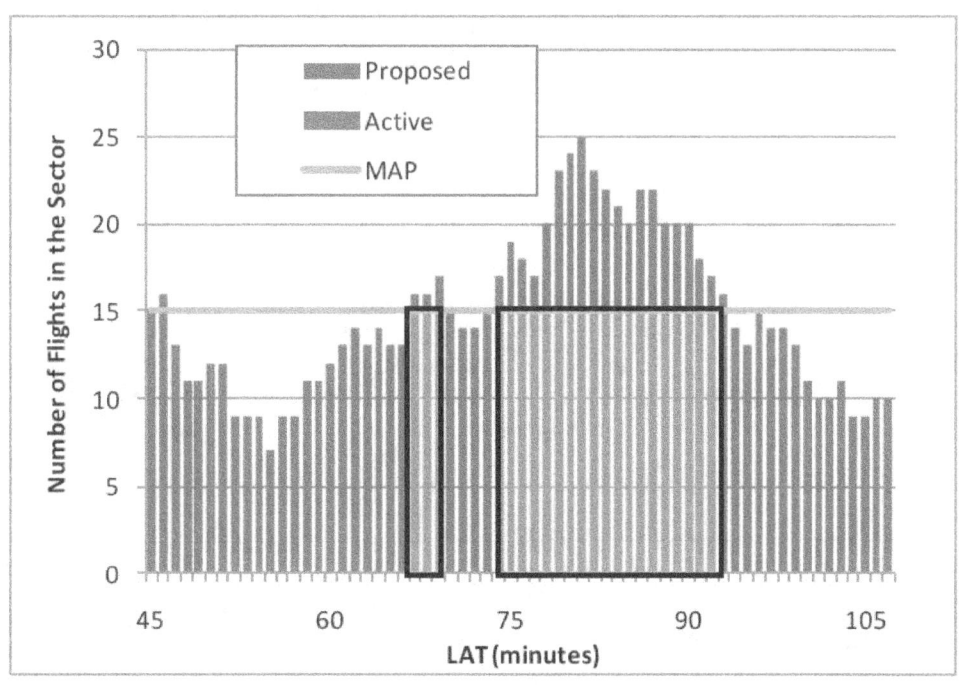

Figure 2-5 Sector Alert Period Identified by the "3 on, 3 off" Alerting Rule

In Figure 2-5, the first alerted period lasts only 3 minutes from the 67th to the 69th minutes (see yellow highlighted area), and the second alerted period lasts 20 minutes from the 74th to the 93rd minutes. The total number of alerted minutes in those two intervals is equal to 23. (Recall that in current TFMS, the entire 60-minute interval in Figure 2-4 is alerted.) In Figure 2-5 the alert periods fell into the time periods where the vast majority of one-minute sector demands exceeded the sector MAP, while the single overloaded minute at LAT=46 substantially separated from the next overloaded interval by normal minutes was ignored by the rule for alerting the sector.

The next example, shown on Figure 2-6, illustrates the "3 on, 5 off" alerting pattern. In comparison with the "3 on, 3 off" pattern, this one increased the number of consecutive normal minutes sufficient for resetting alerts from 3 to 5 minutes. As a result, the two alerted periods from Figure 2-5 were merged into a single and longer alerted period of 27 minutes.

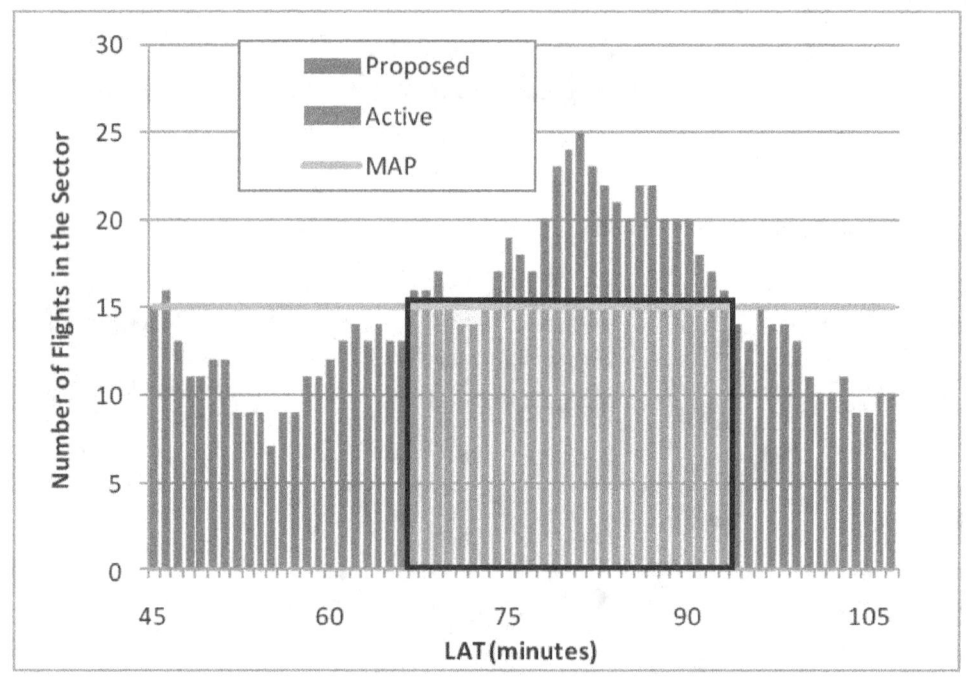

Figure 2-6 Sector Alert Period Identified by the "3 on, 5 off" Alerting Rule

Consider another alerting pattern "5 on, 3 off", in which, in comparison with the previous example of "3 on, 3 off", the number of overloaded minutes sufficient for triggering an alert increased from 3 to 5. Figure 2-7 shows that this pattern provided a single alerted period of 20 minutes. The increase of "on" parameter from 3 to 5 in the alerting pattern, while having the same "off" parameter, reduced the duration of alerted period from 23 to 20 minutes.

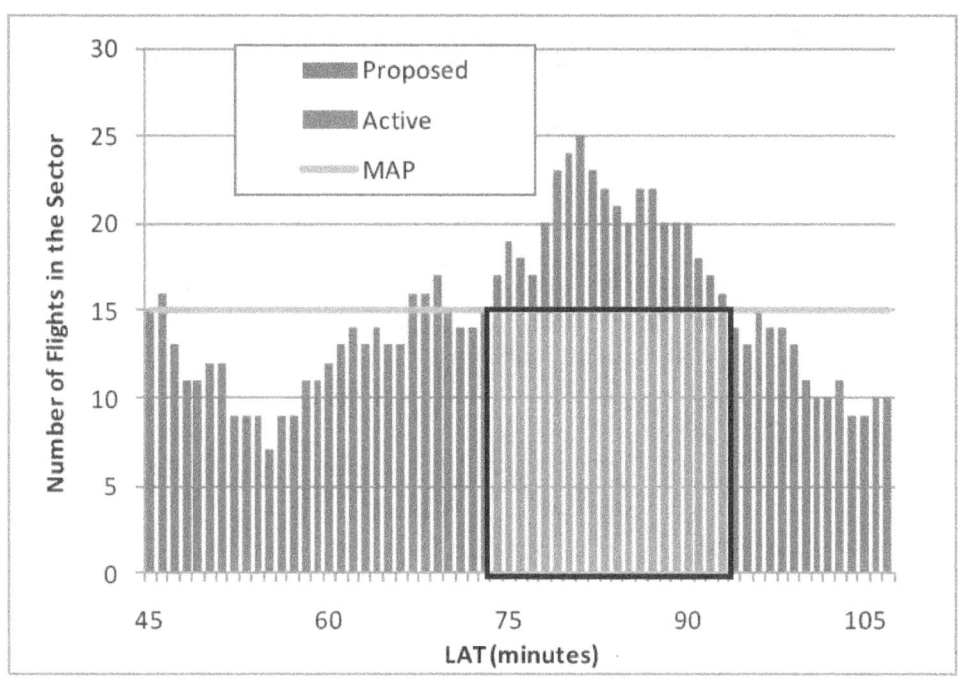

Figure 2-7 Sector Alert Period Identified by the "5 on, 3 off" Alerting Rule

If we increase the "off" parameter from 3 to 5, and apply the "5 on, 5 off" pattern to the same demand, the alerting period will increase from 20 to 27 minutes (see Figure 2-8). In other words, increasing the number of consecutive normal minutes sufficient for resetting alerts with the same "on" minutes in the pattern caused in this example an increase in duration of alerted period. Coincidentally, the alerted periods in Figure 2-6 (the "3 on, 5 off" case) and in Figure 2-8 are the same.

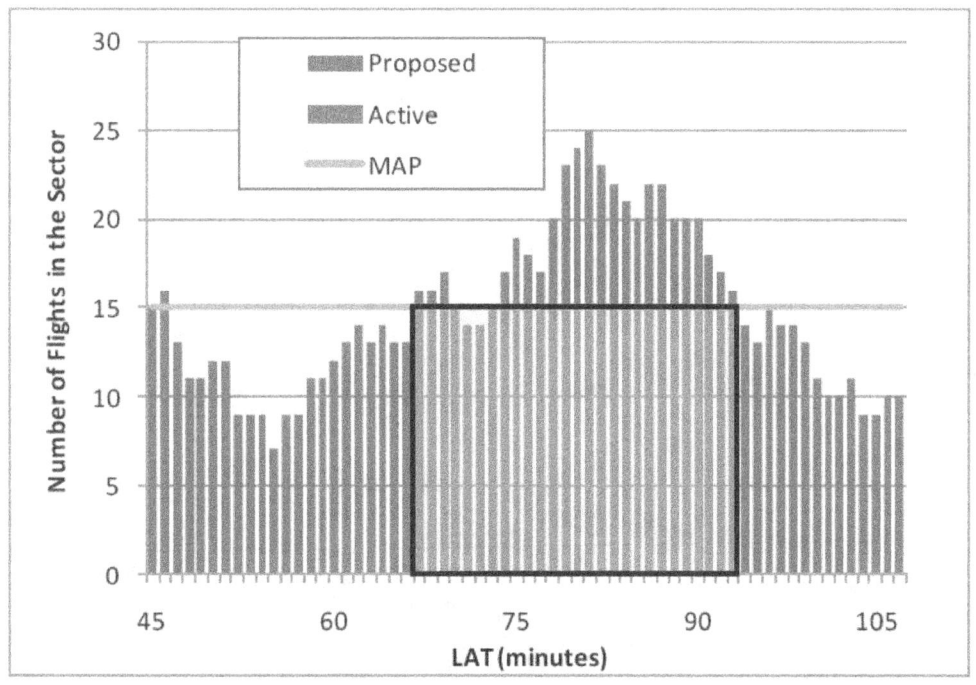

Figure 2-8 Sector Alert Period Identified by the "5 on, 5 off" Alerting Rule

The above examples illustrated some tendencies in changing alerted intervals caused by varying parameters of demand patterns for alerting sectors. Intuitively, it is expected that increasing the minimum number of overloaded minutes sufficient for triggering an alert (parameter *a*) would make it harder (at least not easier) to trigger alerts, and, hence, could cause the reduction in alerted periods. It is also expected that increasing the minimum number of consecutive normal minutes sufficient for resetting alerts (parameter *b*) would make it harder (at least not easier) to end an alert, and, hence would result in increasing (or, at least, not decreasing) periods of alerts. It is also clear that alerted periods identified by the patterns depend on the actual profiles of predicted sector demand. Therefore in order to extract some general tendencies in how parameters of the patters affect sector alert periods, we will need statistical analysis based on much larger sets of traffic demand data than the sets used in this Section for illustrative purposes only. The results of analysis of much larger sets of data are presented in Section 4.

3. Probabilistic Traffic Demand and Alerts

The sector demand patterns discussed in the previous section included deterministic TFMS predictions of one-minute demand counts. However, the FAA NextGen program calls for decision making using probabilistic information. Probabilistic demand predictions acknowledge the inherent uncertainty in demand. Even as flight trajectory predictions improve, our July 2008 report has indicated there will still be substantial uncertainty in the aggregate demand predictions. Our previous research has also indicated that probabilistic predictions are less volatile than the currently used deterministic predictions.

In TFMS, deterministic predictions of one-minute sector demand is based on aggregation of those flights which are expected to be in the sector during the minute of interest. This aggregation does not consider inaccuracy in ETA predictions for individual flights. Analysis has shown [5] that although the error of time-in-sector predictions is small, the standard deviations of error of ETA prediction is larger, typically between 4 and 20 minutes depending on flight's status and look-ahead times. This is a very important factor that may significantly affect accuracy of sector demand predictions and, hence, should be taken into consideration via a probabilistic approach.

In this section, we will give an overview of our previous research on probabilistic representation of one-minute sector demand count predictions.

An analytical method for probabilistic prediction of one-minute sector demand is presented in the Volpe Center's 2010 report [6]. This new method translates probabilistic characteristics of uncertainty in the predictions of arrival times of individual flights into the probabilistic characteristics of uncertainty in aggregate one-minute sector demand counts.

These characteristics depend on both deterministic traffic predictions and parameters of errors in predicting times of flights' sector boundary crossings. The following factors are taken into consideration for determining probabilistic characteristics of one-minute sector demand:

1. Status of flights in TFMS demand predictions: number of active and proposed flights in one-minute counts

2. Distribution of errors in predictions of sector entry times for individual flights depending on flight status and look-ahead time (predictions are significantly more accurate for active flights than for proposed ones, and predictions are somewhat more accurate for shorter look-ahead times)

3. Estimated time in sector (traversing a sector) for individual flights.

Based on this, the mean and standard deviation is derived for both the number of flights *entering* a sector during a given minute and the number of flights *present* in a sector during a given minute. This last derivation provides the basis for a probabilistic prediction of sector demand, with both an expected value of demand and a confidence interval around the expected value.

The main constructive results that provide the analytical ways for calculating characteristics of sector demand predictions are:

1. The expected value of one-minute demand counts is equal of a weighted average of deterministic TFMS predictions within a moving time window that includes a one-minute interval of interest and several preceding and following minutes. The width of time window depends on distribution of errors in predicting times for individual flights: more accurate ETA predictions lead to narrower time windows. The weight coefficients are separately calculated for active and proposed components of TFMS predictions; the widths of time windows can be different for active and proposed fractions of demand.

2. The variances (the square of standard deviations) of one-minute demand predictions are also weighted sums of TFMS predicted one-minute counts within moving time windows with the widths of windows and the weight coefficients calculated differently for active and proposed flights.

The detailed description of the background and computational techniques for probabilistic prediction of one-minute sector demand can be found in the report [6].

To illustrate the probabilistic representation of sector demand, we will use the empirical distributions of errors in predicting a flight's arrival time into a sector separately for active and proposed flights (see Figure 3-1). The distributions were estimated from TFMS data for LAT = 45 min.

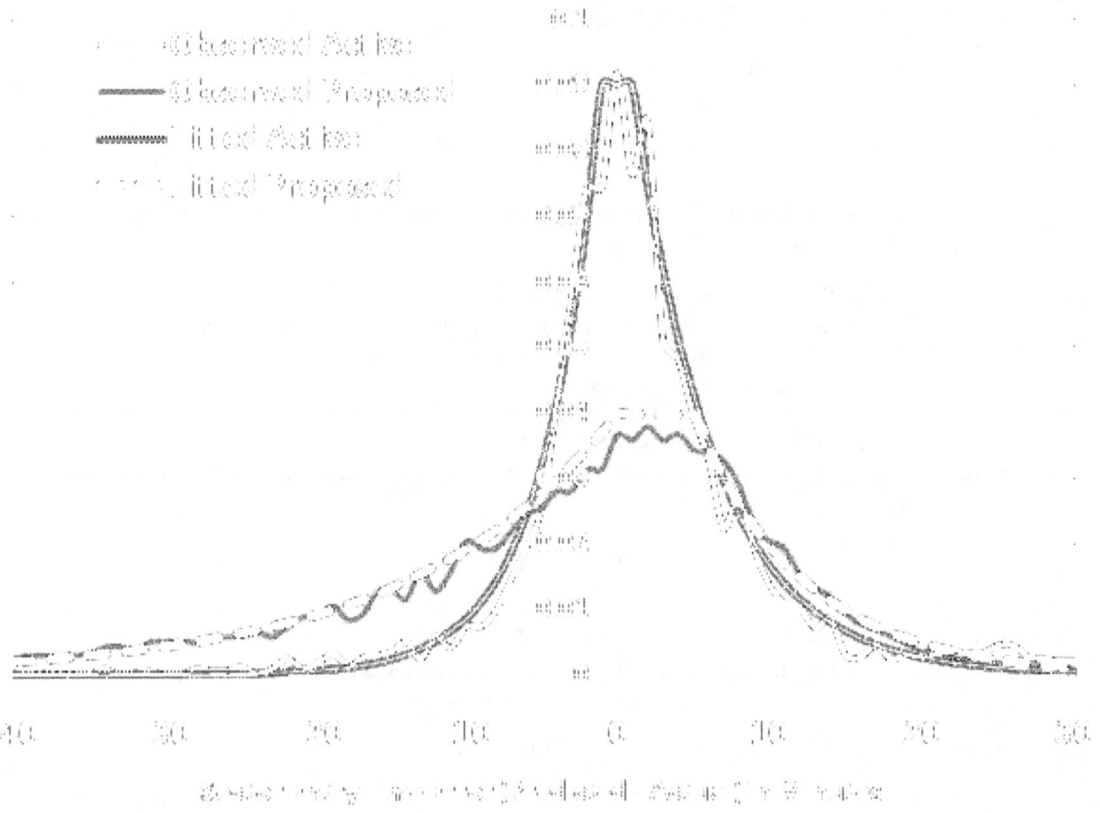

Figure 3-1 Distributions of Errors in Sector Entry Time Predictions and their Approximations

Figure 3-1 shows that the prediction errors have asymmetric distributions: the distribution for proposed flights has heavier left-hand tails while the distribution for active flights has heavier right-hand tails. This kind of behavior of distribution functions reflects the fact that the proposed flights are more likely to arrive at the sector later than earlier while the active flights are more likely to arrive earlier than later relatively to their ETAs.

For the distributions on Figure 3-1, the standard deviations of prediction errors for active and proposed flights are approximately 4 and 15 min, respectively.

The empirical distribution curves were fitted to modified Laplace (double exponential) functions shown in Figure 3-1, that have the following characteristics:

- An exponential curve on the left side

- A small section near 0 with Uniform distribution
- An exponential curve on the right side.

These distribution functions were used for the expected, weighted average demand calculation.

To illustrate how the weighted average function behaves, the weight coefficients from Figure 3-1 were applied to some simple flight demand distributions. In Example 1 (Figure 3-2), assume that the predicted deterministic traffic demand in a sector for several consecutive minutes is equal to 8 flights at each minute, except for a spike that rises to 13 flights for a single minute at LAT = 45. If these flights are all active flights, the expected one-minute demand is determined by averaging deterministic one-minute predictions using the weights given by the green line (Fitted Active) in Figure 3-1. The weighted average curve is the line labeled "If active" in figure 3-2. If they are proposed flights, the sliding time window for averaging follows the dashed purple line (Fitted Proposed) in Figure 3-1. The corresponding weighted average curve is the line labeled "If proposed" In Figure 3-2. For a mix of active and proposed flights, the weighted average curve would fall somewhere between the two curves in the illustration.

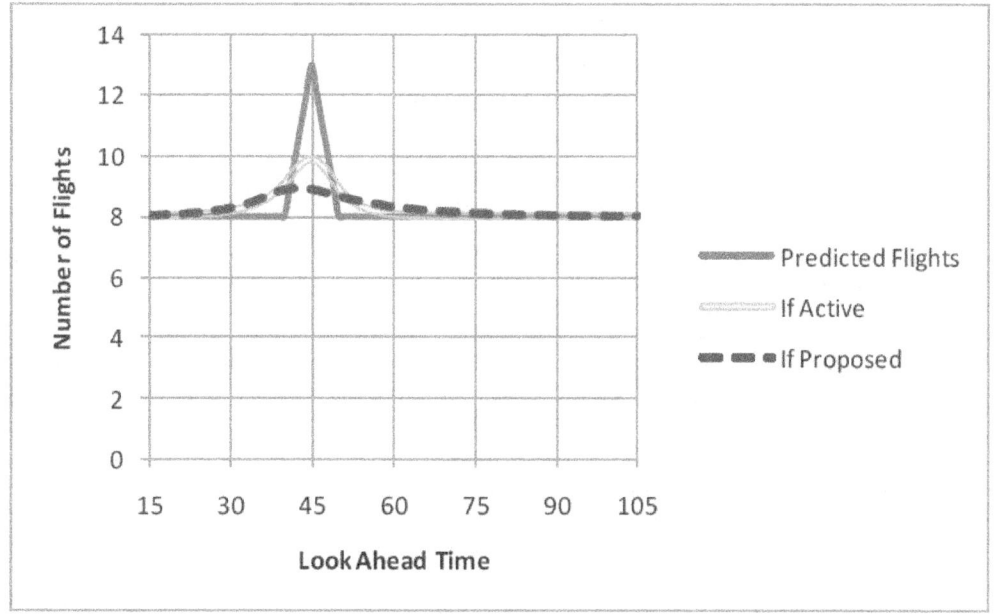

Figure 3-2 Weighted average in Example (1)

Example 2 (Figure 3-3) is similar to Example 1, except that the spike in flights is prolonged, with the 13 flights occurring for 5 minutes rather than 1 minute.

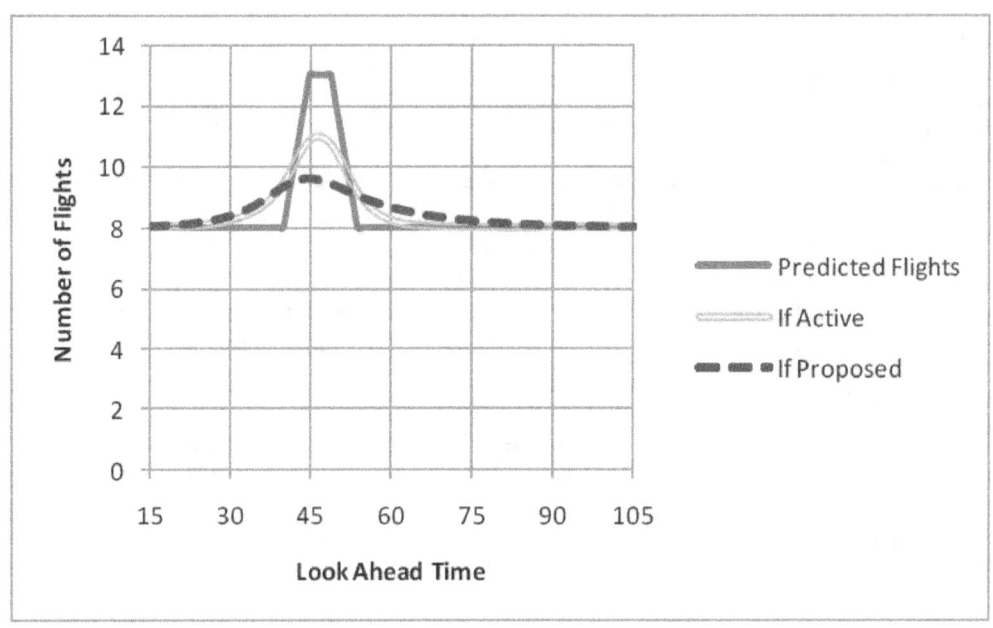

Figure 3-3 Weighted average in Example (2)

In both examples, the local jumps in deterministic sector demand predictions were transformed by weighted averaging into smoother but wider hills with sharper and shorter hills for active flights.

Probabilistic demand predictions are typically presented as uncertainty bands around an expected (mean) value, which was represented by the weighted average in the previous examples. Figure 3-4 illustrates a hypothetical steadily increasing weighted average prediction (the thick purple line labeled "Mean"), compared to a MAP of 15.

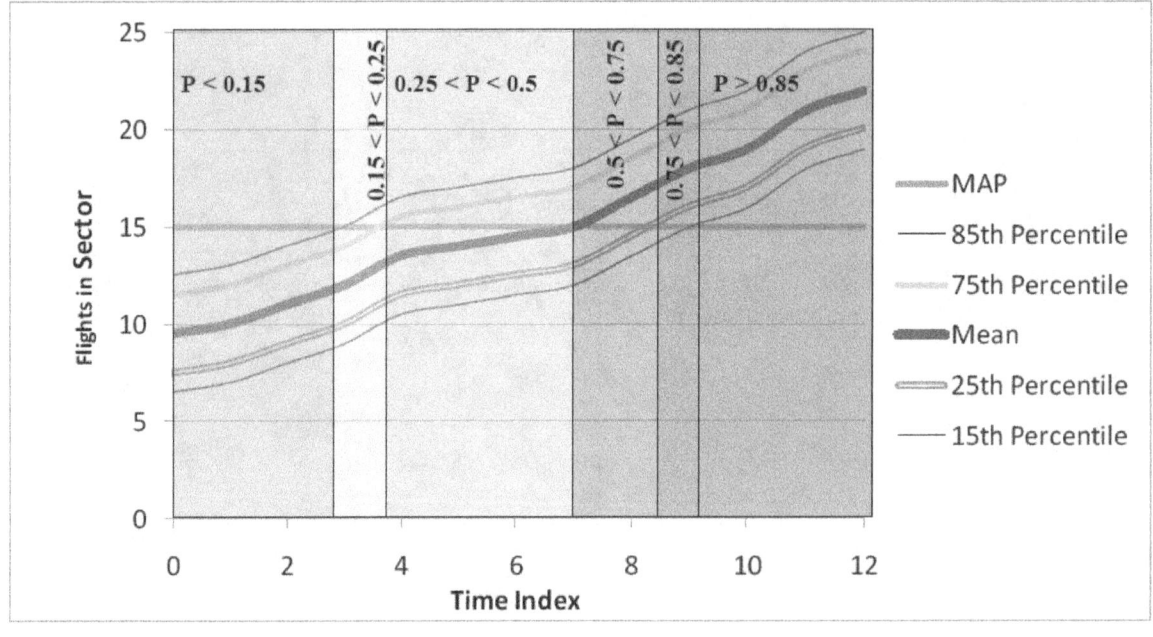

Figure 3-4 Probabilistic Prediction with Uncertainty Bands

In Figure 3-4, the uncertainty band around the mean is between the 25th and 75th percentiles, with auxiliary uncertainty bands (the thin lines) shown at 15th and 85th percentiles.

- In the left portion of the plot (green area), all of the uncertainty bands are less than the MAP. The probability that the number of flights will exceed MAP is less than 15%
- In the yellow area, the MAP falls between the 85th percentile and 75th percentile bands. Thus, the probability that the number of flights will exceed MAP is between 15 and 25%.
- In the orange area, the MAP falls between the 75th percentile and mean bands. Thus, the probability that the number of flights will exceed MAP is between 25 and 50%.
- In the first red area, the MAP falls between the mean and 25th percentile bands. Thus, the probability that the number of flights will exceed MAP is between 50 and 75%.
- In the second red area, the MAP falls between the 25th and 15th percentile bands. Thus, the probability that the number of flights will exceed MAP is between 75 and 85%.
- In the third red area, all of the uncertainty bands are higher than the MAP. The probability that the number of flights will exceed MAP is at least 85%.

In Figure 3-5, the red central area is where the center of the probability band (50%) exceeds the MAP. The orange areas are where the probability of exceeding the MAP is between 25 and 50%. The yellow areas are where the probability of exceeding the MAP is between 15 and 25%. The green area is where the probability of exceeding the MAP is less than 15%.

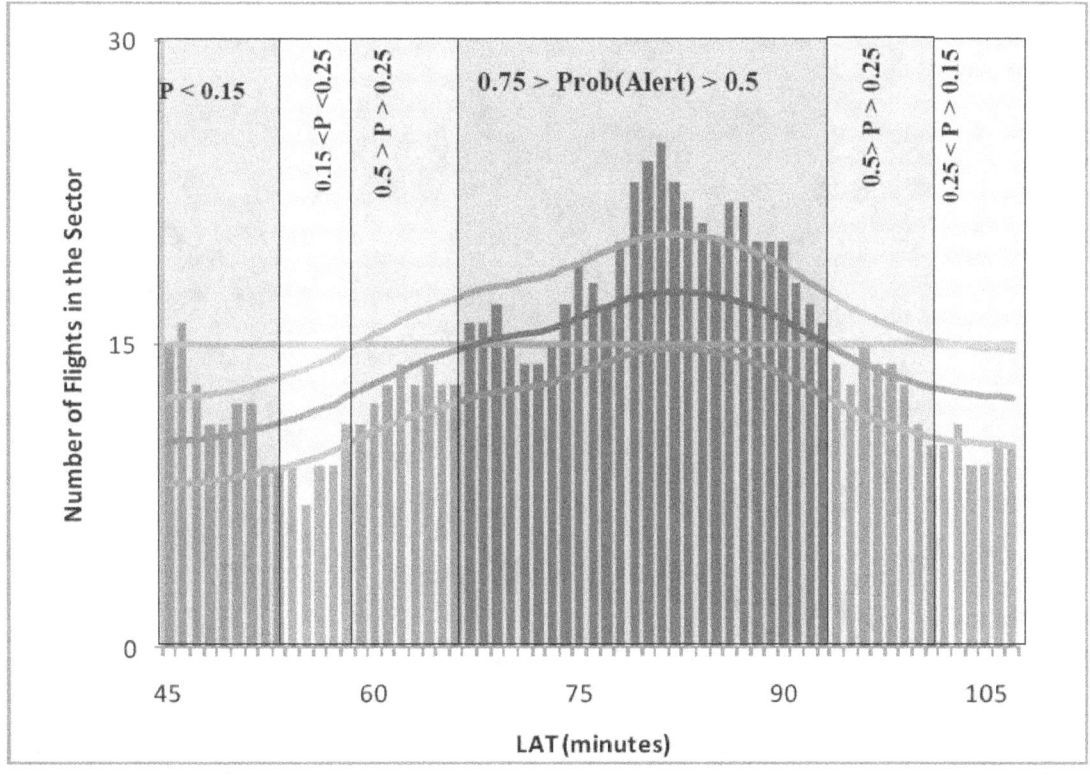

Figure 3-5 Example of Probabilistic Demand Prediction with Uncertainty Band

The advantage of a probabilistic prediction is that it gives a TFM specialist a sense of a likelihood of potential congestion and its severity (both magnitude and duration) that would help him/her in a decision-making process on triggering a TMI: whether to trigger it right now or wait and see.

The issue is how to fit probabilistic ideas into the current workflow, which is based on a simple yes-no alert paradigm. One possible approach is to simply compare the probabilistic expected value (weighted average) to the MAP. This corresponds to the central red area in Figure 3-5. Another possibility is to use a probability other than 0.5 (50%); an alert would be declared whenever the probability of exceeding the MAP exceeds some specified percent. (This approach could also be extended to combine both the patterns described in Section 2, and the probabilistic approach described here.)

4. Analysis of TFMS One-minute Demand Predictions for En Route Sectors

In this section, historical TFMS data is used to illustrate various patterns of one-minute demand that can be considered for sector alerts and how the pattern's parameters affect the temporal characteristics of sector alerts. In addition to the patterns, the weighted average (probabilistic expected value, as discussed in Section 3) of several consecutive one-minute demand predictions within a sliding time window around a one-minute interval of interest was considered as an expected sector demand. The weighted average was also tested for sector alerts and compared with demand patterns in terms of sector alert properties.

The following patterns of one-minute sector demand were considered:

- 3 on, 3 off[1]
- 3 on, 5 off
- 3 on, 8 off
- 5 on, 3 off
- 5 on, 5 off
- 5 on, 8 off
- 8 on, 3 off
- 8 on, 5 off
- 8 on, 8 off

The summary of statistical analyses of various demand patterns applied to alerting potential congestion in en route sectors are presented in Tables 4-1 through 4-3, and in Figure 4-1 and 4-2. The analyzed data was collected from TFMS for

- 16 sectors on April 28, 2010[2], with a total of 24.480 sector-minutes observed
- 14 sectors on April 29, 2010[3], with a total of 4,920 sector-minutes observed
- 11 sectors for several days in April 2009[4], with a total of 781,080 sector-minutes observed.

The tables make it possible to compare various alerting rules with current TFMS and with each other in terms of both the number of alerted minutes (duration) and the number of alerted periods.

---

[1] Section 2, starting on page 4, explains this "on, off" terminology for alert patterns.

[2] ZBW09, ZBW20, ZBW38, ZBW46, ZBW47, ZMP11, ZMP12, ZMP18, ZMP42, ZNY10, ZNY34, ZNY39, ZNY42, ZNY55, ZNY56, ZNY68

[3] ZAB58, ZAB67, ZAB68, ZAB90, ZAB94, ZAB95, ZBW09, ZBW20, ZBW46, ZMP11, ZNY34, ZNY42, ZNY55, ZNY56

[4] ZBW02, ZBW09, ZBW17, ZBW20, ZID83, ZID96, ZLC06, ZMP20, ZOB67, ZOB77, ZTL43. The dates, all in 2009, were April 10 - 12, April 14 - 16, April 18 - 19, and April 24 – 27.

Table 4-1 Alerted Minutes

	Minutes > MAP	Current Measure	3 on 3 off	3 on 5 off	3 on 8 off	5 on 3 off	5 on 5 off	5 on 8 off	8 on 3 off	8 on 5 off	8 on 8 off	Weighted Average
28-Apr	822	3195	795	856	922	634	703	759	433	509	558	335
29-Apr	223	735	208	223	265	183	206	251	142	175	219	29
2009	3320	13680	3004	3153	3309	2309	2407	2527	1500	1577	1731	696
Total	4365	17610	4007	4232	4496	3126	3316	3537	2075	2261	2508	1060

The titles of the columns in those tables are as follows:

- Minutes > MAP (Minutes above MAP). Total number of minutes where the predicted number of flights is above the MAP threshold.
- "Current Measure" corresponds to the results obtained under sector alerting rule in current TFMS. Since the current measure provides alerts in 15-minute blocks, the number of alerted minutes is a multiple of 15.
- "3 on 3 off" through "8 on 8 off" indicate demand patterns used for alerting sectors.
- Weighted average indicates that the expected traffic demand was calculated via a weighted average and then compared to the MAP. The weight coefficients were determined separately for active and proposed flights.

Table 4-1 shows that both demand patterns and weighted average of one-minute sector demands, applied for alerting sectors, significantly reduced the total duration of sector alerts in comparison with the current TFMS Monitor/Alert. This effect was demonstrated on both single-day data (28 and 29 April, 2010) and several days of data of 2009. For example, according to the 2009 row of the table, the TFMS Monitor/Alert identified 13680 minutes of total alert whereas the demand patterns provided total alerted periods ranging from 1500 to 3309 minutes, which are between 4 and 9 times shorter than under current TFMS Monitor/Alert. Weighted average provided an even greater reduction in total alerted periods. Similarly, April 28 data showed that the demand patterns reduced the total TFMS alert periods by factors of between 3 and 7.

Table 4-1 also clearly demonstrates the trends in the relationship between durations of alerted periods and parameters of demand patterns. In all cases in table 4-1,

- Total duration of alerts increases with increasing "off" parameter under the same "on" parameter. For example, in Apr. 28 row, with increasing "off" parameter from 3 to 5 and 8 under "on" parameter equal to 3, the duration of alerts increases from 795 to 856 and 922, respectively.
- Total duration of alerts decreases with increasing "on" parameter under the same "off" parameter. For example, in Apr. 28 row, with increasing "on" parameter from 3 to 5 and 8 under the same "off" parameter equal to 5, the duration of alerts decreases from 856 to 703 and 509 minutes, respectively.

Figure 4-1 gives a graphical representation of the "Total" row of Table 4-1.

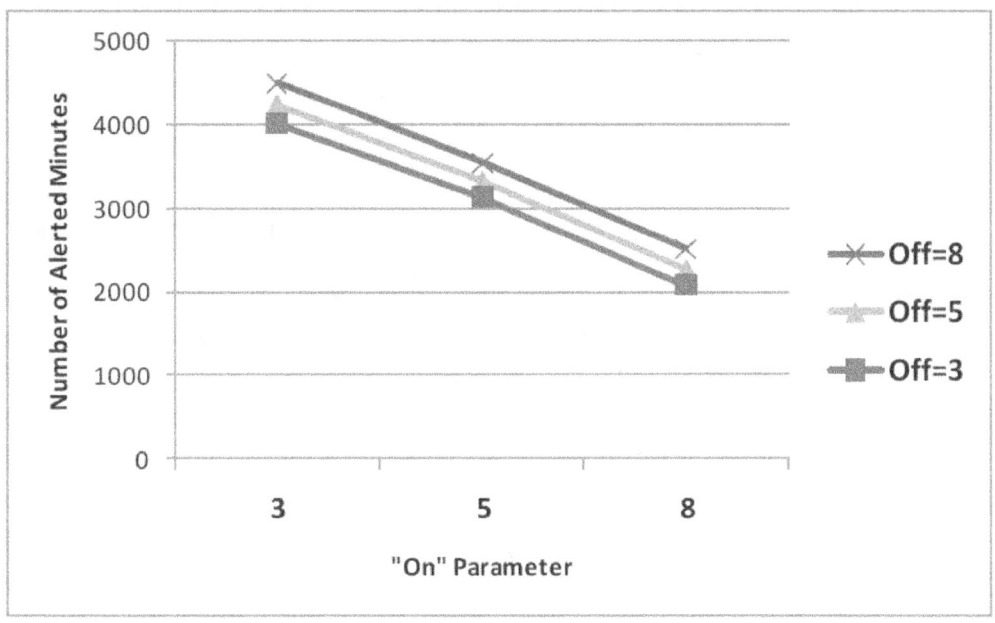

Figure 4-1 Number of Alerted Minutes for the "Pattern" Alerts

Table 4-1 and Figure 4-1 show a clear relationship between the "on" and "off" parameters and the number of alerted minutes:

- As the "on" parameter increases, it becomes harder to turn alerts on because this increases the minimum number of overloaded minutes needed to trigger an alert. The downward slopes of the lines of Figure 4-1 illustrate this trend. Each line corresponds to a single "off" parameter, and shows the change in the total alerted minutes as the "on" parameter increases.

- As the "off" parameter increases, it becomes harder to turn alerts off. As a result, total alerted periods will become longer with a higher "off" parameter. The locations of the three lines in Figure 4-1, which correspond to the three "off" parameters, illustrate this trend.

Table 4-2 shows the number of alerted periods identified by various demand patterns. Using demand patterns as well as weighted average of predicted demand for sector alert identification significantly reduced the number of alerts in comparison with the current TFMS Monitor/Alert.

Table 4-2 and Figure 4-2 show a somewhat different relationship between parameters of demand patterns and the number of alerts. As it is demonstrated in Table 4-2, the "on" parameter makes a big difference (fewer alerts with a higher "on" parameter), but, under the same "on" parameter, the "off" parameter makes no significant difference. Figure 4-2 illustrates this effect as the three lines in the figure practically lie on top of each other.

Table 4-2 Number of Alerts

	Current Measure	3 on 3 off	3 on 5 off	3 on 8 off	5 on 3 off	5 on 5 off	5 on 8 off	8 on 3 off	8 on 5 off	8 on 8 off	Weighted Average
28-Apr	123	96	95	94	56	58	57	26	30	29	13
29-Apr	30	26	24	23	19	19	19	11	13	14	2
2009	630	396	400	396	224	225	220	98	100	100	49
Total	783	518	519	513	299	302	296	135	143	143	64

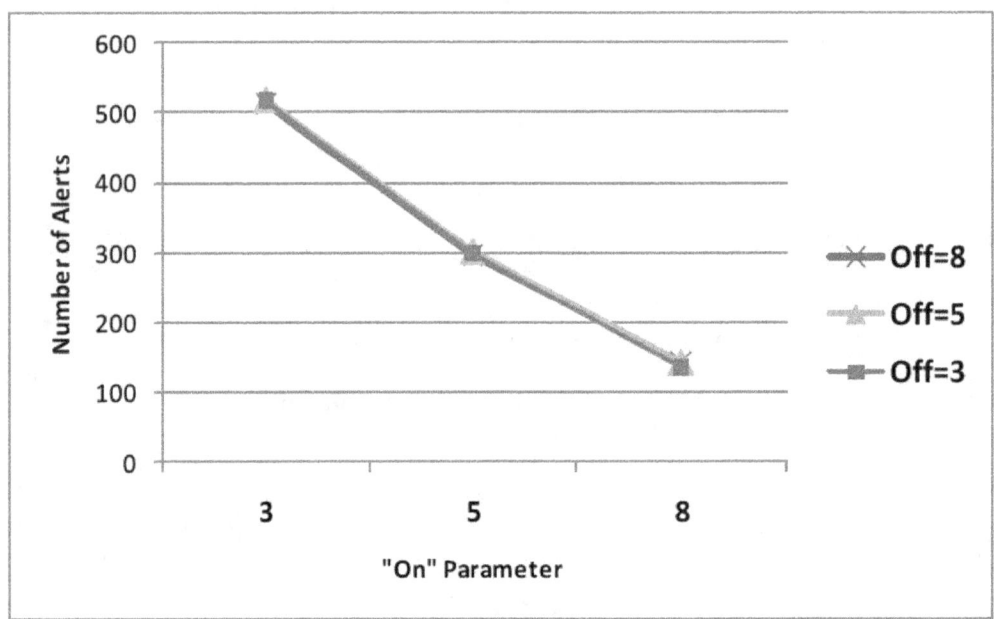

Figure 4-2 Number of Alerts for the "Pattern" Alerts

Finally, one can divide the values in Table 4-1 by the corresponding values in Table 4-2 to obtain an average number of minutes per alert shown in Table 4-3. Again, demand patterns, along with weighted average, significantly reduced the average duration of alerts in comparison with the current TFMS Monitor/Alert (except Apr.28 case where the weighted average and current TFMS have the same average alert duration, but weighted average dramatically reduced the total number of alerts). It is worth noting that, similar to the number of alerts, the change of "on" parameter has greater impact on average duration of alert than the change of "off" parameter under the same value of "on" parameter. Table 4-3 also illustrates that higher parameters of demand patterns result in longer alerts.

Table 4-3 Average Minutes Per Alert

	Current Measure	3 on 3 off	3 on 5 off	3 on 8 off	5 on 3 off	5 on 5 off	5 on 8 off	8 on 3 off	8 on 5 off	8 on 8 off	Weighted Average
28-Apr	26	8	9	10	11	12	13	17	17	19	26
29-Apr	25	8	9	12	10	11	13	13	13	16	15
2009	22	8	8	8	10	11	11	15	16	17	14
Overall	22	8	8	9	10	11	12	15	16	18	17

The next few tables and figures present a drill-down by sector for 28 April 2010, for LAT ranging from 15 minute to 135 minutes.

Table 4-4 Number of Minutes Alerted by Sector, Data from 4/28/2010

Sector	Minutes > MAP	Current Measure	3 on 3 off	3 on 5 off	3 on 8 off	5 on 3 off	5 on 5 off	5 on 8 off	8 on 3 off	8 on 5 off	8 on 8 off	Weighted Average
ZBW09	15	45	20	20	20	9	9	9				
ZBW20	73	435	62	62	71	45	45	45	20	20	20	
ZBW38	5	45	5	5	5							
ZBW46	108	330	105	111	111	95	101	101	76	89	89	165
ZBW47	17	75	5	14	14	5	5	5				
ZMP11	12	75	12	12	12							
ZMP12	51	105	49	49	49	49	49	49	49	49	49	16
ZMP18	37	120	34	38	38	17	28	28	17	17	17	
ZMP42	5	45	3	3	3							
ZNY10	165	270	182	188	210	170	176	198	146	152	174	94
ZNY34	138	645	147	164	171	128	149	156	94	111	118	51
ZNY39	9	90	3	3	13							
ZNY42	126	480	119	138	156	86	111	138	31	71	91	9
ZNY55	20	195	9	9	9							
ZNY56	35	195	37	37	37	30	30	30				
ZNY68	6	45	3	3	3							
Grand Total	822	3195	795	856	922	634	703	759	433	509	558	335

Table 4-4 shows the numbers of minutes alerted identified by various sector alert patterns including weighted average and summarized for all sectors. The total TFMS data set consisted of more than 24 thousand one-minute demand predictions for 16 sectors. Of those 24 thousand minutes, 822 minutes were overloaded, i.e., predicted demand exceeded sectors' MAPs. Current TFMS sector alerting rules provided total 3195 minutes of sector alerts, which is 4 times greater than the total number of overloaded minutes. All alerting patterns considered significantly reduced the total minutes of sector alerts ranging from 433 to 922, roughly 3 to 7 times smaller than the total minutes of TFMS sector alerts.

The table confirmed a general effect of alerting patterns on duration of alerts that was discussed earlier. However, the quantitative effects vary by sectors. For ZBW20 and ZNY34, all alerting patterns significantly reduced the duration of alerts in comparison with TFMS: the reduction for ZBW20 ranged from 84% (for 3 on, 8 off pattern) to 95% (for 8 on, 3 off pattern); the reduction for ZNY34 ranged from 73% (for 3 on, 8 off pattern) to 85% (for 8 on, 3 off pattern). A smaller reduction in the total alert duration was obtained at ZNY10 that ranged from 29% (for 3 on, 5 off pattern) to 46% (for 8 on, 3 off pattern).

Weighted average significantly smoothed the demand profile and significantly reduced the total duration of alerts. The weighted average provides the expected values of demand, which is a part of the probabilistic representation of traffic demand. However, in addition to expected demand, a true probabilistic representation requires a corridor of uncertainty around expected values that covers a confidence area with a specific probability ranges.

Table 4-5 shows the numbers of alerts. Table 4-6 shows the average minutes per alert, calculated as Number of minutes alerted (from Table 4-4) divided by Number of alerts (from Table 4-5).

Table 4-5 Number of Alerts, Data from 4/28/2010

Sector	Current Measure	3 on 3 off	3 on 5 off	3 on 8 off	5 on 3 off	5 on 5 off	5 on 8 off	8 on 3 off	8 on 5 off	8 on 8 off	Weighted Average
ZBW09	3	3	3	3	1	1	1				
ZBW20	18	11	11	12	6	6	6	2	2	2	
ZBW38	3	1	1	1							
ZBW46	11	7	7	7	5	5	5	2	3	3	2
ZBW47	3	1	2	2	1	1	1				
ZMP11	4	3	3	3							
ZMP12	4	3	3	3	3	3	3	3	3	3	2
ZMP18	8	7	6	6	2	3	3	2	2	2	
ZMP42	2	1	1	1							
ZNY10	9	14	14	12	11	11	9	8	8	6	6
ZNY34	13	15	15	15	11	12	12	6	7	7	1
ZNY39	5	1	1	2							
ZNY42	16	20	19	18	12	12	13	3	5	6	2
ZNY55	11	2	2	2							
ZNY56	10	6	6	6	4	4	4				
ZNY68	3	1	1	1							
Grand Total	123	96	95	94	56	58	57	26	30	29	13

Table 4-5 shows that demand patterns for alerting sectors work differently in different sectors: at the majority of sectors considered the demand patterns consistently reduce the number of alerted periods in comparison with the current TFMS Monitor/Alert (e.g., ZBW20, ZBW46, ZMP18), while at three ZNY sectors, ZNY10, ZNY34 and ZNY42, the number of alerted periods were increased under some demand patterns: at ZNY34 and ZNY42 under the "3 on" patterns, and at ZNY10 under both "3 on" and "5 on" patterns. Increasing in the number of alerts is not necessarily a negative effect if we, for instance, compare a *single* one-hour alert in current TFMS caused by, say, from 2 to 5 overloaded minutes in each of four consecutive 15-minute intervals with *three* alerted periods during the same one-hour with the total duration of 30 minutes identified by demand patterns. This effect is clearly demonstrated from both Table 4-4 and 4-5: for those three sectors with increased number of alerted periods, the same demand patterns, which caused increase in the numbers, reduced the total duration of alerted periods by 73% - 77% in ZNY34, by 67% - 75% in ZNY42, and by 22% - 33% in ZNY10 under "3 on" patterns.

Table 4-6 Average Minutes per Alert, Data from 4/28/2010

Sector	Current Measure	3 on 3 off	3 on 5 off	3 on 8 off	5 on 3 off	5 on 5 off	5 on 8 off	8 on 3 off	8 on 5 off	8 on 8 off	Weighted Average
ZBW09	15	7	7	7	9	9	9				
ZBW20	24	6	6	6	8	8	8	10	10	10	
ZBW38	15	5	5	5							
ZBW46	30	15	16	16	19	20	20	38	30	30	83
ZBW47	25	5	7	7	5	5	5				
ZMP11	19	4	4	4							
ZMP12	26	16	16	16	16	16	16	16	16	16	8
ZMP18	15	5	6	6	9	9	9	9	9	9	
ZMP42	23	3	3	3							
ZNY10	30	13	13	18	15	16	22	18	19	29	16
ZNY34	50	10	11	11	12	12	13	16	16	17	51
ZNY39	18	3	3	7							
ZNY42	30	6	7	9	7	9	11	10	14	15	5
ZNY55	18	5	5	5							
ZNY56	20	6	6	6	8	8	8				
ZNY68	15	3	3	3							
Grand Total	26	8	9	10	11	12	13	17	17	19	26

The average minutes per alert (Table 4-6) behaves as expected. Under the current measure, the average duration of alert is long, because even a single minute above the threshold will trigger a 15-minute alert. As the parameters for the pattern measures increase, the average minutes per alert becomes longer. This is for several reasons:

- The "on" parameter sets a lower bound on the duration of an alert. For example, if at least three overloaded minutes are needed, the duration of the alert will never be less than 3 minutes
- A higher "on" parameter makes it harder to turn alerts on, eliminating many short alerts
- A higher "off" parameter makes it harder to turn alerts off, thus resulting in longer alerts. It may also result in several alerts being combined into one alert (recall Figure 2-5 and Figure 2-6).

However, averages can often mask interesting trends. The next few figures, taken from 4/28/2010 data, show the distributions of the durations of alerts for various alerting patterns. They are normalized to

100%. For example,

Figure 4-3 presents the histogram for the "Current Measure", which has an average of 26 minutes per alert from Table 4-6. 62% of those alerts are in the 15-19 minute bucket (actually, 15 minutes, since the number of minutes alerted for each Current Measure alert is a multiple of 15). 25% are in the 30-39 minute bucket (actually, 30 minutes), 7% are in the 40-49 minute bucket (actually, 45 minutes), and the remaining 7% are 60 minutes or longer.

Figure 4-3 Current TFMS Measure: Histogram of Alert Duration

Figure 4-4 shows the histograms for the pattern 3 on 3 off, 3 on 5 off, and 3 on 8 off alerts. Figure 4-5 shows the histograms for the pattern 5 on 3 off, 5 on 5 off, and 5 on 8 off alerts. Figure 4-6 shows the histograms for the pattern 8 on 3 off, 8 on 5 off, and 8 on 8 off alerts. Finally, Figure 4-7 shows the histogram of durations for the weighted average alerts.

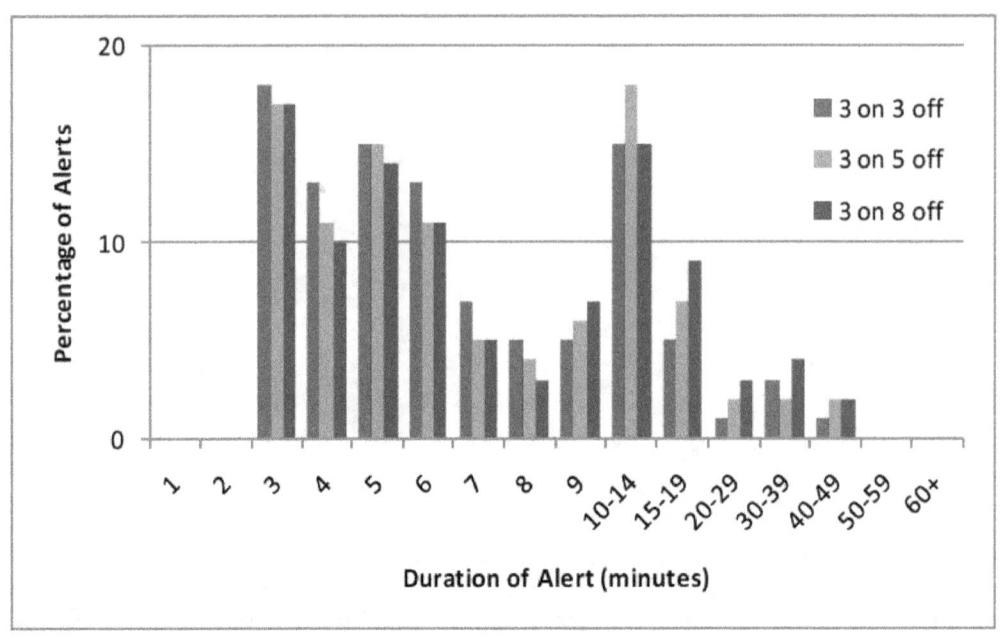

Figure 4-4 Pattern "3 on" Alerts: Histograms of Alert Durations

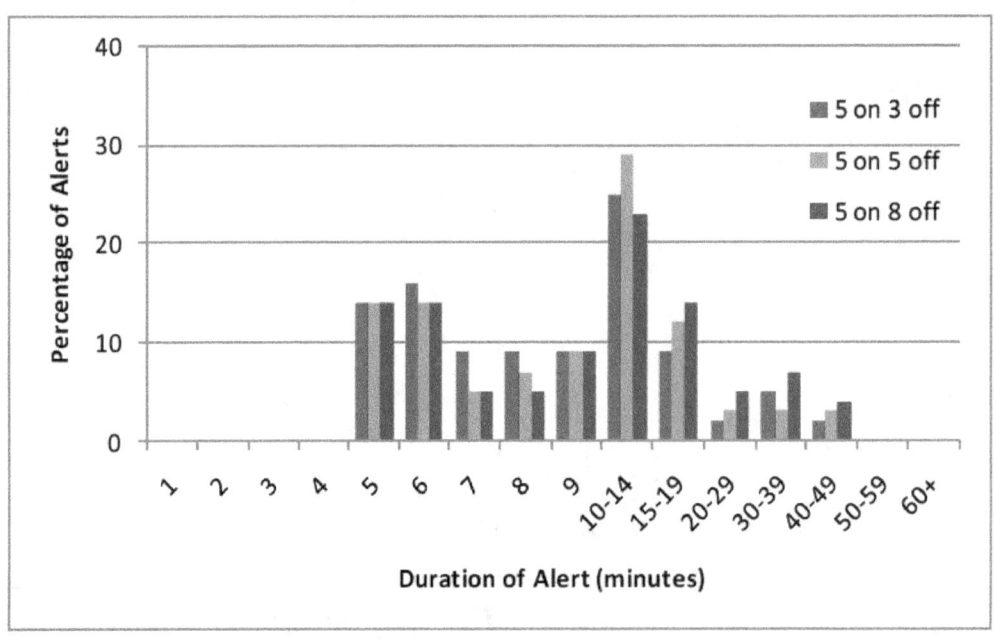

Figure 4-5 Pattern "5 on" Alerts: Histograms of Alert Durations

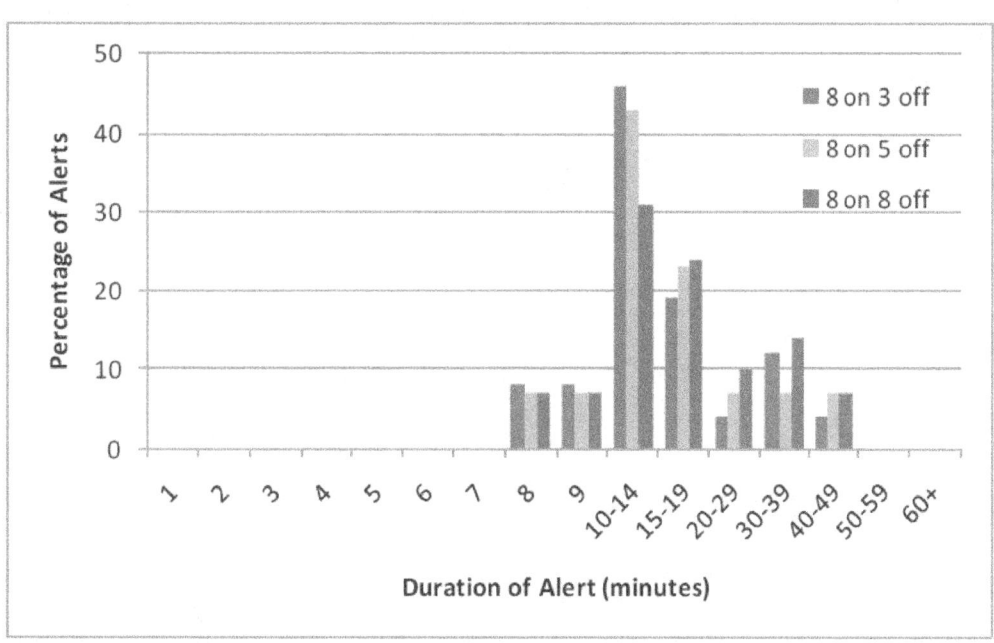

Figure 4-6 Pattern "8 on" Alerts: Histograms of Alert Durations

Comparing Figure 4-4, Figure 4-5, and Figure 4-6, one can see a number of trends. First, as the "on" parameter increases, the durations become longer (they move to the right). This is primarily because, as discussed earlier, the "on" parameter sets a hard lower bound on the duration of an alert. Second, looking within each of the figures, one can see an increase in duration as the "off" parameter increases. The purple bars (corresponding to the "8 off" parameter) are higher for the longer durations. Finally, the "3 on" and "5 on" patterns tend to produce shorter alert durations than those under current rules

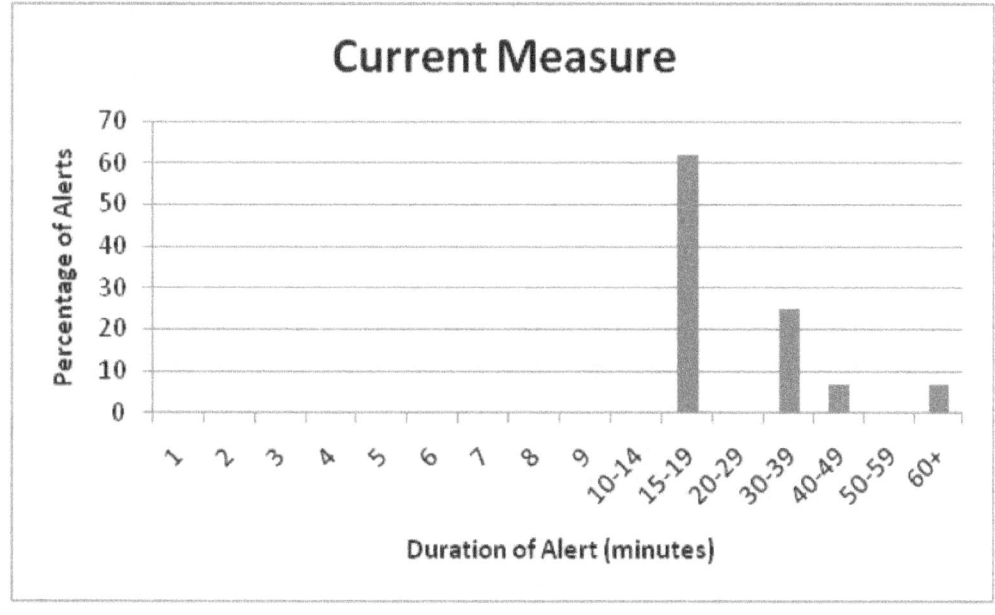

Figure 4-3).

Figure 4-7 shows a histogram of duration for the weighted average alert rule. Here, there is a wide variety of durations. Approximately 20% of the alerts are less than 10 minutes (under weighted average, there is nothing to preclude even a one-minute alert duration, should the average slightly exceed the MAP for a single minute). However, most of the alerts are in the 10 – 30 minute range.

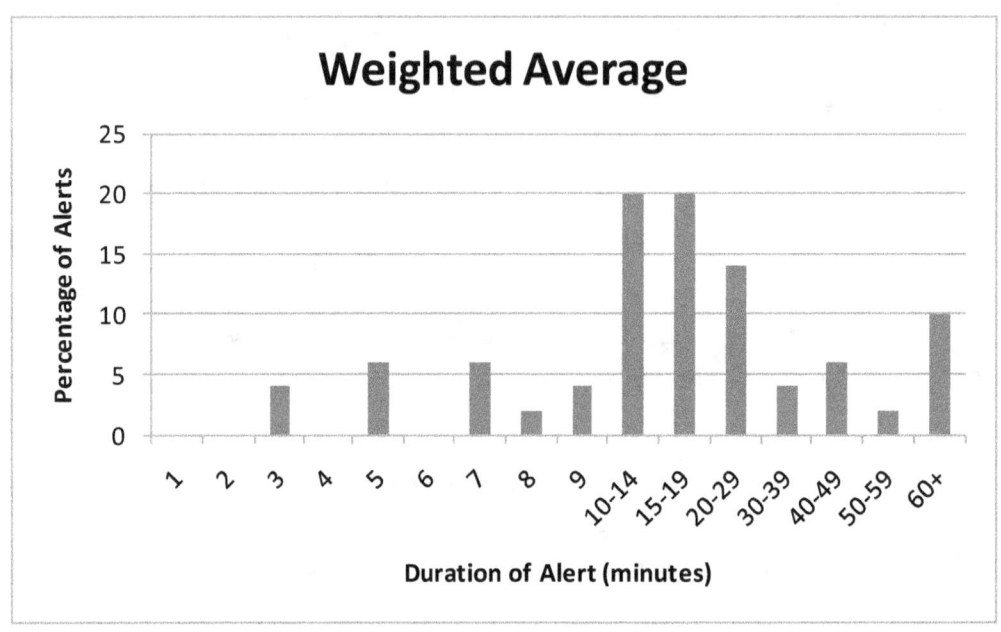

Figure 4-7 Sector Alert by Weighted Average of Demand: Histogram of Alert Durations

5. Summary and Next Steps

This report presents the latest results of research conducted at the Volpe Center on improving air traffic demand predictions and enhancing Traffic Flow Management System (TFMS) Monitor/Alert function for identifying potential congestion at National Airspace System (NAS) elements, such as airports, sectors and fixes. This report is devoted to en route sectors.

The goal of this study is to develop a methodology that would allow for comparative analysis of various options for enhancing TFMS Monitor/Alert functionality in identifying potential congestion in airspace. This study addresses the following questions:

- What benefits and to what extent do the proposed new rules for measuring sector alerts provide for improving current TFMS Monitor/Alert functionality in predicting potential congestion in sectors?

- How many alerts are created under various alerting rules? What is their duration?

- How sensitive are sector alert periods to parameters of demand patterns?
    - at the same sector
    - at various sectors

- How do probabilistic predictions of one-minute sector demands translate into probabilistic Monitor/Alert?

This study is based on analysis of TFMS demand prediction data at several en route sectors during several days of 2009 and 2010. We created a methodology to process real-time sector demand prediction data to identify sector alerts. We also created a constructive algorithm for probabilistic predictions of sector one-minute traffic demand, and for translating those predictions to probabilistic Monitor/Alert.

The major results of this study appear below.

1. The aggregated results for all considered sectors showed that:

- Demand patterns and weighted average of one-minute sector demand predictions significantly reduced the number of alerted periods in comparison with current TFMS Monitor/Alert

- Demand patterns and weighted averages of one-minute sector demands used for identification of potential sector alerts significantly reduced the total duration of alerted periods in comparison with current TFMS Monitor/ Alert

- Demand patterns reduced not only the total duration but also the average number of minutes per alert (or average duration of a single alert)

- Weighted average of sector demand did not noticeably reduce the average number of minutes per alert but the weighted average significantly reduced the total duration of alerts in comparison with current TFMS Monitor/Alert

- As the "on" parameter of demand patterns increases, it becomes harder to turn alerts on because of the increased minimum duration of an alerted period. There are fewer alerts and the total duration of alerts decreases

- As the "off" parameter of demand patterns increases, it becomes harder to turn an alert off. As a result, the total duration of alerted periods tend to be higher with a higher "off" parameter

- The number of alerted periods is significantly more sensitive to the "on" parameter (under various "off" parameters) than to the "of parameter (under the same "on" parameter) of demand patterns

2. **The results for individual sectors showed that:**

   - When demand patterns are used, different sectors react differently in terms of number of identified alerts relative to current TFMS: the number may decrease in some sectors and increase in other sectors under the same parameters of demand patterns. However, in all cases considered, the total duration of alerts substantially decreased under demand patterns regardless of whether the number of alerted periods decreased or increased in comparison with current TFMS

   - Demand patterns reduced total duration of alerted periods at each sector in comparison with current TFMS Monitor/Alert

   - The magnitude of reduction in total duration of alerts relativel to current TFMS varies by sector. Among the sectors considered, the maximum reduction was within the range between 84% and 95% at ZBW20, and minimum reduction was within the range between 29% and 46% at ZNY10

3. **The results for probabilistic sector alerts showed that:**

   - Expected one-minute traffic demand predictions were estimated via a weighted average of several consecutive TFMS predictions. The weighted average was based on probabilistic considerations, namely the observed error distribution of sector entry times for active and proposed flights.

   - We estimated a range of uncertainty in the 1-minute demand predictions, and constructed uncertainty bands around the weighted average, also based on prediction errors for active and proposed flights.

   - Based on weighted average and range of uncertainty, the probabilities of alerts were determined at various times, which would provide further information for TFM decision-making.

This study offers to the FAA alternatives to the current TFMS way of alerting sectors. These alternatives are of interest because the one-minute peak demand prediction in 15-minute (currently used in TFMS for sector alerts) is not accurate; it is unstable and does not reflect controllers' workload. The patterns of one-minute demand with properly selected parameters are able to reflect tendencies in sector one-minute demands in terms of both magnitude of demand and cohesiveness of potentially overloaded minutes that would lead to more reliable identification of sector alerts. Probabilistic approach to sector alerts by smoothing one-minute demand counts via weighted average within a sliding time window together with a confidence area around the average makes it possible to take into account uncertainty in sector demand predictions and provides another option for more realistic Monitor/Alert for en route sectors.

The parameters of demand patterns for alerting sectors or the probabilities for declaring alerts can be updated by TFM specialists to reflect the changes in air traffic, changes in perception of conditions for alerting sectors or changes in number of controllers that in turn may lead to changes in the workload thresholds.

The following questions need the expert opinion of TFM specialists:

- Whether the concept of demand patterns is a good idea for improving sector Monitor/Alert and whether the patterns adequately reflect the perception of alerts by TFM specialists and controllers' workload

- Whether weighted average and probabilistic rules for sector alerts are preferable to TFM specialists

- Whether parameters of demand patterns for sector alerts should be the same for all sectors or they should be sector-specific and how to select these parameters

In closing, it should be stressed that here are two directions that the Monitor/Alert capability can evolve to take uncertainty into account and to deal with the shortcomings of the present approach:

- Stay within the current framework in which an alert is either on or off. The method proposed in this report is to use the pattern of demand, rather than the demand for a single minute, to determine if a sector should be alerted. This pattern can be defined using the "**a** on, **b** off" scheme defined in this report or by using a weighted average of minute-by-minute demand in an interval surrounding the interval of interest.
- Break with the current framework by using a fully probabilistic approach. In this approach an alert is declared if the probability of the MAP being exceeded is above a specified percentage.

To be determined is which, if either, of the approaches is deemed suitable for operational deployment.

6. References

1. FAA's NextGen Implementation Plan, FAA, March 2010

2. NextGen Mid-Term Concept of Operations for the National Airspace System, FAA, April 30, 2010

3. Smith, Scott and Eugene Gilbo, "Analysis of Uncertainty in ETMS Aggregate Demand Predictions," Report Number VNTSC-ATMS-05-05, 2005.

4. Gilbo, Eugene and Rick Oiesen, "A New Method for Determining a Sector Alert," Report Number VNTSC-TFM-08-11, 2008.

5. Gilbo, Eugene and Scott Smith, "Characterization of Uncertainty in ETMS Flight Events Predictions and its Effect on Traffic Demand Predictions," Report Number VNTSC-TFM-08-09, 2008.

6. Gilbo, Eugene and Scott Smith, "Probabilistic Predictions of Traffic Demand for En Route Sectors Based on Individual Flight Data," Report Number VNTSC-TFM-10-01, 2010

www.ingramcontent.com/pod-product-compliance
Lightning Source LLC
Chambersburg PA
CBHW081805170526
45167CB00008B/3334